BEYOND THE VISIBLE

The Search for Dark Matter

BY:-
AMAZIR AMANATH
(DIP. IN ASTRONOMY)
(CEO - NEBULA NOVA ACADEMY)

Beyond The Visible

The Search for Dark Matter

By
Amazir Amanath
(Diploma in Celestial Navigation)
(CEO - Nebula Nova Academy)

| Beyond the Visible: The Search for Dark Matter | © Amazir Amanath | 2023 |

Writer's Words

As a 15-year-old aspiring astronaut and astronomer, I am thrilled to share this exploration of dark matter with you. My journey into the cosmos began with a fascination for the stars and an eagerness to understand the universe's greatest mysteries. Writing this book has been a dream come true, allowing me to delve deeply into one of the most enigmatic and compelling topics in modern science.

While I am still in the early stages of my scientific career, my passion for astronomy and space exploration drives me to seek out knowledge and share it with others. The quest to understand dark matter, a fundamental yet elusive component of our universe, is a testament to the power of curiosity and perseverance. I hope that this book not only provides valuable insights but also inspires other young minds to pursue their own dreams in science and exploration.

Thank you for joining me on this journey through the cosmos. I look forward to continuing my studies and contributing to our collective understanding of the universe.

With curiosity and enthusiasm,
Amazir Amanath
(Diploma in Celestial Navigation)

| Beyond the Visible: The Search for Dark Matter | © Amazir Amanath | 2023 |

Contents

Front Page i

Writer's Words ii

Contents iii

Chapter 1: Introduction to Dark Matter 1

Chapter 2: The Discovery of Dark Matter 6

Chapter 3: Properties and Characteristics 10

Chapter 4: Theoretical Frameworks 15

Chapter 5: Detection Methods and Experiments 20

Chapter 6: Astrophysical Evidence 25

Chapter 7: Cosmological Implications 31

Chapter 8: Challenges and Controversies 35

Chapter 9: Dark Matter in Popular Culture 38

Chapter 10: Conclusion and Future Directions 42

Reference & Glossary 47

Chapter 1: Introduction to Dark Matter

What is Dark Matter?

Dark matter is one of the most enigmatic components of the universe. Unlike ordinary matter, which makes up stars, planets, and living organisms, dark matter does not interact with electromagnetic forces. This means it does not emit, absorb, or reflect light, rendering it invisible to telescopes and other instruments that detect electromagnetic radiation. Despite its invisibility, dark matter makes up about 27% of the universe's mass-energy content, vastly overshadowing the mere 5% constituted by visible matter.

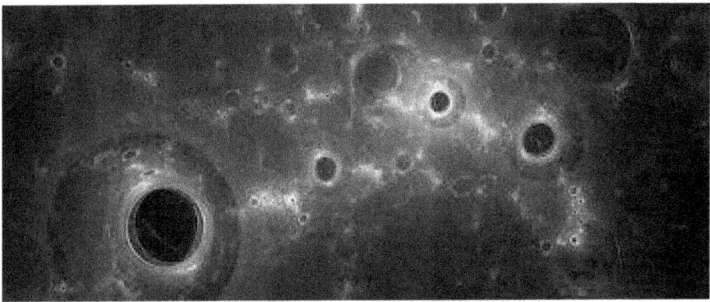

The term "dark matter" was coined to describe this mysterious substance, which we infer exists due to its gravitational effects on visible matter. Its presence is suggested by various astronomical and cosmological observations, such as the rotation curves of galaxies, the motion of galaxy clusters, and the large-scale structure of the universe.

Historical Background

The concept of dark matter emerged from observations in the early 20th century. Swiss astronomer Fritz Zwicky first proposed the existence of dark matter in 1933. While studying the Coma Cluster of galaxies, he noticed that the galaxies were moving at speeds too high to be accounted for by the visible matter alone. Zwicky suggested that a substantial amount of unseen mass was present, which he referred to as "dunkle Materie" or dark matter.

However, it wasn't until the 1970s that dark matter gained widespread attention. Vera Rubin's work on the rotation curves of spiral galaxies provided compelling evidence for dark matter. Rubin observed that the stars in the outer regions of galaxies were moving much faster than would be expected based solely on the visible mass. This discrepancy led to the realization that additional unseen mass was influencing the rotation speeds of these stars.

Why Study Dark Matter?

Understanding dark matter is crucial for several reasons:

1. Cosmology: Dark matter plays a fundamental role in the formation and evolution of the universe. It affects the large-scale structure of the cosmos, influencing how galaxies and galaxy clusters form and evolve. Without dark matter, our current models of the universe would be incomplete and inaccurate.

2. Physics: Dark matter could provide insights into new physics beyond the Standard Model of particle physics. It may be composed of particles that interact with ordinary matter in ways other than through gravity, offering clues about fundamental forces and particles.

3. Technology: The quest to detect dark matter has driven the development of advanced technologies in particle detection and observational astronomy. These technologies often find applications beyond their initial scientific goals, contributing to advancements in various fields.

The Nature of Dark Matter

Dark matter is hypothesized to be composed of particles that do not interact with electromagnetic forces. This makes it invisible to current telescopes and observational techniques that rely on detecting light. Instead, dark matter's presence is inferred from its gravitational effects on visible matter.

Several candidates have been proposed for dark matter particles. Among the leading candidates are:

- WIMPs (Weakly Interacting Massive Particles): Predicted by theories such as supersymmetry, WIMPs are heavy particles that interact through weak nuclear forces.
- Axions: Lightweight particles proposed to address certain problems in quantum chromodynamics.
- MACHOs (Massive Compact Halo Objects): Ordinary matter in the form of objects like black holes or neutron stars, though these cannot account for all dark matter.

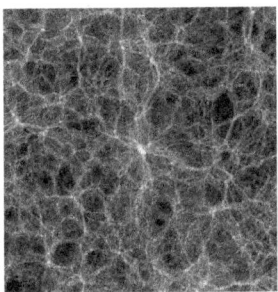

Despite these hypotheses, no dark matter particles have been directly detected to date. Researchers continue to explore these and other candidates through various experimental and observational methods.

How Dark Matter is Detected

Dark matter is detected indirectly through its gravitational influence. Key methods of detecting dark matter include:

1. Gravitational Lensing: This technique involves observing the bending of light from distant objects due to the gravitational field of dark matter. By studying the distortions in the light, scientists can map the distribution of dark matter.

2. Galactic Rotation Curves: By measuring the rotational speeds of stars in galaxies, scientists can infer the presence and distribution of dark matter. The faster-than-expected rotation speeds in the outer regions of galaxies suggest the presence of dark matter.

3. Cosmic Microwave Background (CMB): The CMB provides a snapshot of the early universe. Analyzing its temperature fluctuations and polarization helps scientists understand the distribution of dark matter in the early cosmos.

4. Direct Detection Experiments: These experiments aim to detect dark matter particles interacting with regular matter. They are typically conducted deep underground to minimize interference from cosmic rays.

5. Indirect Detection: Researchers look for the byproducts of dark matter interactions, such as gamma rays or neutrinos, which can provide evidence for dark matter.

Dark Matter's Role in the Universe

Dark matter influences the large-scale structure of the universe. It forms halos around galaxies and galaxy clusters, providing the gravitational scaffolding necessary for their formation and stability. Without dark matter, the formation of galaxies and the large-scale structure observed today would be impossible.

Dark matter's gravitational effects also impact the behavior of galaxy clusters and the distribution of galaxies across the universe. The observed distribution of galaxies and the large-scale structure align with theoretical models that include dark matter.

The Future of Dark Matter Research

Despite significant progress, many questions about dark matter remain unanswered. Future research aims to:

- Improve Detection Methods: New technologies and experimental techniques are being developed to enhance our ability to detect dark matter directly and indirectly.
- Refine Theoretical Models: Researchers continue to explore various theoretical models to explain dark matter and its properties.
- Investigate Dark Matter's Role: Understanding dark matter's influence on cosmic structures and its role in the universe's evolution remains a key focus.

Conclusion

Dark matter remains one of the greatest mysteries in modern astrophysics and cosmology. Its study is essential for a complete understanding of the universe, from the formation of galaxies to the large-scale structure we observe today. As research continues, the quest to uncover the true nature of dark matter promises to reveal profound insights into the fundamental workings of the cosmos.

Chapter 2: The Discovery of Dark Matter

Early Observations and Anomalies

The story of dark matter begins in the early 20th century with observations that hinted at the existence of unseen mass in the universe. In the 1930s, Swiss astronomer Fritz Zwicky was studying the Coma Cluster, a group of galaxies approximately 320 million light-years from Earth. Zwicky applied the virial theorem, a principle in classical mechanics that relates the average kinetic energy of a system to its average potential energy, to calculate the mass of the cluster.

To his surprise, Zwicky found that the visible mass of the galaxies was insufficient to account for the observed gravitational effects. The galaxies were moving much faster than expected based on their visible matter alone, suggesting that there must be a substantial amount of unseen mass providing the necessary gravitational pull to keep the cluster bound together. Zwicky called this unseen mass "dunkle Materie," or dark matter. Although his findings were initially met with skepticism, they laid the groundwork for future discoveries.

Further anomalies were observed in the 1950s and 1960s. Astronomers noted that the distribution of mass in galaxies did not match the distribution of light. For instance, the outer regions of spiral galaxies rotated at unexpectedly high speeds. According to Newtonian mechanics, stars and gas at the periphery of a galaxy should move more slowly than those near the center, where most of the visible mass is concentrated. However, this was not observed, suggesting the presence of additional, invisible mass.

Vera Rubin and Galactic Rotation Curves

A pivotal moment in the study of dark matter came in the 1970s, with the work of American astronomer Vera Rubin and her colleague Kent Ford. Rubin and Ford conducted detailed studies of the rotational velocities of stars in spiral galaxies, particularly focusing on the Andromeda Galaxy (M31). Using advanced spectroscopic techniques, they measured the Doppler shift of stars, which provided information about their speeds and directions of movement.

Their observations revealed that the rotational velocity of stars remained relatively constant with increasing distance from the galactic center. This finding was unexpected because, based on the visible matter alone, one would anticipate a decline in velocity at greater distances due to weaker gravitational pull. This phenomenon, known as the "flat rotation curve," implied that galaxies contained much more mass than what was visible, likely extending far beyond the observable disk of stars.

Rubin's work was significant not only for its scientific impact but also for its timing. By the 1970s, the scientific community had started to accept the notion of dark matter, thanks in part to advances in technology and observational techniques. Rubin's research provided robust evidence that dark matter was not just a curiosity but a fundamental component of galaxies.

Gravitational Lensing and Evidence

Another compelling line of evidence for dark matter comes from gravitational lensing, a phenomenon predicted by Einstein's theory of general relativity. Gravitational lensing occurs when a massive object, such as a galaxy or galaxy cluster, lies between a distant light source and an observer. The gravity of the intervening object bends the light from the background source, distorting its image and sometimes creating multiple images of the same object.

Gravitational lensing allows astronomers to measure the mass of the lensing object, including both visible and invisible matter. In many cases, the mass inferred from gravitational lensing is significantly greater than the mass estimated from visible matter alone, providing direct evidence for the presence of dark matter. One of the most striking examples of gravitational lensing is the Bullet Cluster, a pair of colliding galaxy clusters. Observations of the Bullet Cluster reveal a separation between the visible matter, primarily in the form of hot gas detected through X-ray emissions, and the total mass distribution inferred from gravitational lensing. The lensing mass, which corresponds to the dark matter, does not align with the visible matter, indicating that dark matter does not interact strongly with ordinary matter or electromagnetic forces.

Contributions of Various Observations

The confirmation of dark matter's existence has been bolstered by a range of astronomical observations and techniques. Beyond galactic rotation curves and gravitational lensing, the Cosmic Microwave Background (CMB) provides another crucial line of evidence. The CMB is the afterglow of the Big Bang, a relic radiation that fills the universe. Tiny fluctuations in the temperature of the CMB reveal information about the universe's composition and structure during its early stages. Data from the CMB, particularly from the Wilkinson Microwave Anisotropy Probe (WMAP) and the Planck satellite, show patterns that can only be explained by the presence of dark matter. These observations indicate that dark matter was crucial in the formation of the first galaxies and large-scale structures in the universe, acting as a gravitational scaffold around which ordinary matter could accumulate.

Implications and Current Understanding

The discovery of dark matter has profound implications for our understanding of the universe. It suggests that the vast majority of matter in the universe is invisible and interacts primarily through gravity. The exact nature of dark matter remains one of the biggest mysteries in cosmology and particle physics. While several candidates, such as Weakly Interacting Massive Particles (WIMPs) and axions, have been proposed, none have been conclusively detected.

Current research focuses on refining our understanding of dark matter through a combination of observational, experimental, and theoretical approaches. Particle detectors, such as those at the Large Hadron Collider (LHC) and underground laboratories, are actively searching for signs of dark matter interactions. At the same time, astronomers continue to use telescopes and satellites to map the distribution of dark matter in the universe, hoping to learn more about its properties and behavior.

Conclusion

The journey from the initial observations of unexplained gravitational effects to the widespread acceptance of dark matter as a fundamental component of the universe has been a remarkable scientific adventure. The discovery of dark matter has reshaped our understanding of the cosmos, revealing a universe that is far more complex and mysterious than previously imagined. As research progresses, the quest to uncover the true nature of dark matter continues, promising new insights into the fundamental workings of the universe.

Chapter 3: Properties and Characteristics

The Nature of Dark Matter

Dark matter is an elusive substance that does not interact with electromagnetic radiation, meaning it neither emits, absorbs, nor reflects light. This characteristic renders it invisible to telescopes and other instruments that rely on detecting light or electromagnetic waves. The primary evidence for dark matter comes from its gravitational effects on visible matter, radiation, and the large-scale structure of the universe.

Composition and Hypothetical Particles

Several hypotheses exist about the nature of dark matter, primarily focused on its composition:

1. Weakly Interacting Massive Particles (WIMPs): WIMPs are a popular candidate for dark matter. They are theorized to interact only through gravity and the weak nuclear force, making them difficult to detect. These particles could have masses ranging from a few GeV to several TeV (giga-electron volts to tera-electron volts), and their weak interactions imply that they rarely collide with ordinary matter, making direct detection challenging.

2. Axions: Axions are hypothetical particles proposed as a solution to the strong CP problem in quantum chromodynamics (QCD). They are extremely light and interact very weakly with ordinary matter. Axions could also be a component of cold dark matter, contributing to the formation of cosmic structures.

3. Sterile Neutrinos: These are a type of neutrino that does not interact via the standard weak nuclear force, unlike regular neutrinos. Sterile neutrinos are heavier and interact only through gravity, making them a viable dark matter candidate.

4. MACHOs (Massive Compact Halo Objects): MACHOs include objects such as black holes, neutron stars, brown dwarfs, and rogue planets. While these can contribute to the overall mass of a galaxy, they cannot account for all dark matter due to their insufficient abundance.

5. Primordial Black Holes: Formed in the early universe, these black holes could make up a portion of dark matter. However, their contribution is debated, as they might not provide a sufficient explanation for the total dark matter content observed.

Cold, Warm, and Hot Dark Matter

Dark matter is also categorized based on the velocity of its particles:

1. Cold Dark Matter (CDM): CDM consists of slow-moving particles. It is the most widely accepted form of dark matter, as it can explain the formation of galaxies and large-scale structures. The "cold" designation refers to its low thermal velocity relative to the speed of light.

2. Warm Dark Matter (WDM): WDM particles have velocities between those of cold and hot dark matter. WDM could explain the distribution of small galaxies and the smoothness of the dark matter distribution on smaller scales.

3. Hot Dark Matter (HDM): HDM consists of fast-moving particles, such as neutrinos with low masses. HDM cannot explain the current structure of the universe alone, as its high velocity would prevent the clumping of matter necessary for galaxy formation.

Distribution and Interaction

Dark matter is not evenly distributed throughout the universe. Instead, it forms clumps and structures that influence the formation and evolution of galaxies and galaxy clusters.

Galactic Halos and Clusters

One of the most significant concentrations of dark matter is found in galactic halos. These halos extend far beyond the visible components of galaxies, providing the necessary gravitational pull to explain the observed rotation curves. In the absence of dark matter, the outer stars in galaxies would not have the observed high velocities.

In galaxy clusters, dark matter plays a crucial role in binding galaxies together. Observations show that the visible matter (stars, gas) in clusters accounts for only a fraction of the total mass. The remaining mass, attributed to dark matter, helps maintain the integrity of the cluster despite the high velocities of individual galaxies within it.

Large-Scale Structure

On cosmic scales, dark matter forms a web-like structure, often referred to as the "cosmic web." This structure is composed of filaments of dark matter connecting galaxy clusters and voids, regions with relatively little matter. The distribution of galaxies follows this web-like pattern, with clusters and superclusters located at the intersections of filaments.

The cosmic web's formation is a consequence of gravitational instabilities in the early universe. Tiny fluctuations in the density of matter, amplified by gravity, led to the formation of the first dark matter halos. These halos attracted ordinary matter, eventually forming stars and galaxies. The presence of dark matter is essential for this process, as it provides the gravitational potential wells needed for baryonic matter to accumulate and cool, leading to star formation.

Interaction with Ordinary Matter

Dark matter's interaction with ordinary matter, or baryonic matter, appears to be limited primarily to gravitational forces. This lack of significant interaction with electromagnetic forces means dark matter does not participate in processes like photon emission or absorption, making it invisible to traditional observational methods.

However, several experiments aim to detect dark matter through other potential interactions:

1. Direct Detection: These experiments seek to observe the collisions between dark matter particles and atomic nuclei. Such interactions are expected to be exceedingly rare due to the weak nature of dark matter's interactions. Detectors

often use ultra-pure materials and are located deep underground to shield them from cosmic rays and other background noise.

2. Indirect Detection: This method involves searching for secondary particles or radiation resulting from dark matter annihilation or decay. For example, the collision of dark matter particles might produce gamma rays, neutrinos, or positrons, which can be detected by various astrophysical observatories.

3. Collider Experiments: Particle accelerators, such as the Large Hadron Collider (LHC), are used to search for dark matter particles by recreating conditions similar to those just after the Big Bang. By colliding particles at high energies, scientists hope to produce dark matter particles or observe missing energy signatures that could indicate their presence.

The Bullet Cluster and Evidence from Collisions

One of the most striking pieces of evidence for dark matter comes from observations of colliding galaxy clusters, such as the Bullet Cluster. In these collisions, the hot gas, which constitutes most of the visible mass, interacts electromagnetically and slows down, while the dark matter components pass through each other without significant interaction.

Gravitational lensing studies of the Bullet Cluster reveal that the gravitational mass (including dark matter) is located separately from the visible mass. This separation provides compelling evidence that dark matter is a distinct component that does not interact strongly with itself or with baryonic matter, except through gravity.

The Role in Cosmic Evolution

Dark matter has played a crucial role in the evolution of the universe from the earliest moments to the present day. Its gravitational influence helped seed the formation of the first structures, leading to the development of galaxies, stars, and ultimately, life. Without dark matter, the universe would be a very different place, possibly devoid of the complex structures we observe today.

The study of dark matter continues to be a vibrant field of research, with implications for cosmology, particle physics, and our broader understanding of the universe. As technology advances and new observational techniques are developed, we hope to unlock more secrets about this mysterious substance and its fundamental properties.

Chapter 4: Theoretical Frameworks

Standard Model Limitations

The Standard Model of particle physics is a well-established theory that describes the fundamental particles and their interactions, excluding gravity. It includes particles like quarks, leptons (such as electrons and neutrinos), gauge bosons (which mediate forces), and the Higgs boson. Despite its success, the Standard Model has significant limitations when it comes to explaining dark matter.

Lack of Dark Matter Candidates

The Standard Model does not account for dark matter. None of the particles within the model have the properties necessary to explain the observed phenomena associated with dark matter, such as its gravitational effects on galaxies and the cosmic microwave background. For instance, neutrinos, which are the only particles within the Standard Model that do not interact electromagnetically, are too light and too fast-moving to account for the dark matter that is needed to form large-scale structures in the universe.

Hierarchy Problem

The hierarchy problem refers to the question of why gravity is so much weaker than the other fundamental forces. In the context of dark matter, this issue is relevant because many proposed dark matter candidates, such as WIMPs, are tied to extensions of the Standard Model that attempt to address the hierarchy problem. The weakness of gravity compared to other forces makes it challenging to detect dark matter particles, as they would primarily interact through gravitational forces.

Fine-Tuning and Naturalness

The Standard Model requires a high degree of fine-tuning to match observations, particularly regarding the mass of the Higgs boson. This fine-tuning problem has led physicists to seek extensions to the Standard Model that can provide a more natural explanation for the observed values of physical constants. Many such extensions also introduce potential dark matter candidates.

Dark Matter Candidates

Several candidates for dark matter have been proposed, most of which involve particles that lie beyond the Standard Model. These candidates can be broadly categorized based on their masses and interaction strengths.

Weakly Interacting Massive Particles (WIMPs)

WIMPs are one of the most studied and promising dark matter candidates. They are hypothetical particles that interact via the weak nuclear force and gravity. Their mass is typically predicted to be in the range of a few GeV to several TeV. WIMPs arise naturally in several extensions of the Standard Model, including:

- Supersymmetry (SUSY): In SUSY, each particle in the Standard Model has a corresponding "superpartner" with different spin properties. The lightest supersymmetric particle (LSP) is stable in many models and can act as a dark matter candidate. The neutralino, a mixture of superpartners of the Higgs boson and gauge bosons, is a well-known WIMP candidate.

- Extra Dimensions: Theories involving extra spatial dimensions predict the existence of particles that appear as WIMPs in our three-dimensional space. These particles could include Kaluza-Klein excitations, which are higher-dimensional analogs of particles in the Standard Model.

Axions

Axions are hypothetical particles that were originally proposed to solve the strong CP problem in quantum chromodynamics (QCD). They are very light, with masses that could range from micro-electron volts (μeV) to milli-electron volts (meV). Axions are cold dark matter candidates because they are produced non-relativistically in the early universe. Unlike WIMPs, axions do not interact via the weak nuclear force but may interact with photons and electrons at extremely low rates.

The Axion Dark Matter Experiment (ADMX) and other experiments aim to detect axions by looking for their conversion into photons in the presence of a magnetic field, a process predicted by the axion-photon interaction.

Sterile Neutrinos

Sterile neutrinos are hypothetical neutrinos that do not interact via the weak nuclear force, unlike the three known types of neutrinos in the Standard Model. They are considered "sterile" because they do not interact except through gravity. Sterile neutrinos could have masses ranging from a few keV to several MeV and are considered warm dark matter candidates. They could explain some of the discrepancies in the observed distribution of galaxies and the cosmic microwave background.

Primordial Black Holes

Primordial black holes (PBHs) are black holes that could have formed in the early universe due to density fluctuations. Unlike stellar black holes, which form from collapsing stars, PBHs could have a wide range of masses, including very small masses. Depending on their mass distribution, PBHs could contribute to the dark matter density. However, constraints from gravitational lensing, cosmic microwave background observations, and other astrophysical phenomena limit their abundance.

Alternative Theories

In addition to these specific dark matter candidates, several alternative theories have been proposed to explain the gravitational effects attributed to dark matter. These theories often involve modifications to the laws of gravity or new forms of matter and energy.

Modified Newtonian Dynamics (MOND)

MOND is an alternative to dark matter that modifies Newton's laws of motion at very low accelerations, typically found in the outer regions of galaxies. MOND suggests that the gravitational force is stronger than predicted by Newtonian gravity at these scales, potentially explaining the flat rotation curves of galaxies without invoking dark matter. However, MOND struggles to explain the observations of the cosmic microwave background and the large-scale structure of the universe as effectively as dark matter models.

Modified Gravity Theories

Several theories propose modifications to General Relativity to account for the effects attributed to dark matter. One such theory is TeVeS (Tensor-Vector-Scalar gravity), which modifies General Relativity by introducing additional fields. These fields can mimic the effects of dark matter on galactic scales but often face challenges in explaining cosmological observations, such as the cosmic microwave background and galaxy cluster dynamics.

Another approach is f(R) gravity, where the gravitational Lagrangian is modified to include a function of the Ricci scalar, R. This modification can lead to extra gravitational forces that could account for the effects usually attributed to dark matter. However, these theories often require fine-tuning and additional assumptions to fit observational data.

Emergent Gravity

Emergent gravity theories suggest that gravity is not a fundamental force but an emergent phenomenon arising from microscopic interactions. These theories often involve concepts from quantum information theory and thermodynamics. One such theory, proposed by Erik Verlinde, posits that the observed gravitational effects attributed to dark matter are due to an emergent, entropic force related to the information content of space. While intriguing, emergent gravity theories are still in the early stages of development and require further exploration to match observational data as accurately as standard dark matter models.

Conclusion

The quest to understand dark matter involves a rich interplay between theory, observation, and experimentation. While the Standard Model of particle physics provides a robust framework for understanding fundamental forces and particles, it falls short of explaining dark matter. As a result, physicists and astronomers have proposed a variety of dark matter candidates and alternative theories. These include WIMPs, axions, sterile neutrinos, and modifications to gravitational theory.

Current and future experiments aim to directly detect dark matter particles or their interaction products, while ongoing astronomical observations continue to refine our understanding of the universe's structure and composition. The continued exploration of dark matter promises to deepen our understanding of the cosmos and potentially reveal new physics beyond the Standard Model.

Chapter 5: Detection Methods and Experiments

The search for dark matter involves various detection methods and experimental approaches, aimed at identifying the elusive particles or observing their indirect effects. The challenge lies in the fact that dark matter interacts weakly, if at all, with ordinary matter and electromagnetic radiation, making it difficult to detect with conventional instruments.

Direct Detection

Direct detection experiments aim to observe dark matter particles interacting with normal matter. The goal is to detect the rare collisions between dark matter particles, typically WIMPs, and atomic nuclei within a detector.

Detection Techniques

1. Cryogenic Detectors: These detectors are cooled to extremely low temperatures to reduce thermal noise and detect the tiny amounts of energy deposited by WIMP collisions. Materials like germanium or silicon are commonly used. When a dark matter particle collides with a nucleus in the detector, it can cause the nucleus to recoil, producing a detectable signal in the form of ionization, phonons (vibrational energy), or scintillation (light).

2. Noble Gas Detectors: These detectors use noble gases like xenon or argon in a liquid state. When a dark matter particle interacts with the nucleus of a noble gas atom, it can produce scintillation light and ionize the gas. The ionization electrons are then drifted by an electric field and detected. The dual-phase technology, where both scintillation and ionization signals are measured, helps distinguish between potential dark matter signals and background noise.

3. Scintillation Detectors: These detectors use materials that emit light (scintillate) when a particle interacts with them. The emitted light is then detected by photomultiplier tubes or other light sensors. Scintillation detectors can be used in conjunction with other detection methods to enhance the sensitivity and discrimination power against background events.

Background Noise and Discrimination

One of the biggest challenges in direct detection is distinguishing potential dark matter signals from background noise, which can arise from natural radioactivity, cosmic rays, and other environmental sources. To minimize background noise, experiments are often conducted deep underground, where they are shielded from cosmic rays, and employ ultra-pure materials and sophisticated data analysis techniques to identify and eliminate spurious signals.

For example, the LUX-ZEPLIN (LZ) experiment, located in the Sanford Underground Research Facility, is one of the leading efforts to detect WIMPs. It uses a large volume of liquid xenon, which is highly sensitive to potential dark matter interactions, and is shielded by rock and water to reduce background radiation.

Indirect Detection

Indirect detection methods seek to observe the secondary products of dark matter interactions, such as gamma rays, neutrinos, or antiparticles. The underlying idea is that dark matter particles, if they are capable of self-annihilation or decay, would produce these standard particles, which can then be detected.

Gamma Rays

Gamma rays are a primary focus in indirect detection because they can travel across vast distances without being significantly deflected or absorbed. Observatories such as the Fermi Gamma-ray Space Telescope and the Cherenkov Telescope Array (CTA) monitor the sky for excess gamma rays that could indicate dark matter annihilation or decay, particularly in regions expected to contain high concentrations of dark matter, such as the Galactic Center, dwarf spheroidal galaxies, or galaxy clusters.

Neutrinos

Neutrinos, being neutral and weakly interacting, are challenging to detect but can provide valuable information about dark matter. Neutrino observatories like IceCube in Antarctica detect the tiny flashes of light produced when neutrinos interact with ice or water. An excess of neutrinos from specific regions in the sky could signal dark matter annihilation processes.

Cosmic Rays

Cosmic rays, which consist of high-energy particles, can also provide indirect evidence for dark matter. For instance, an excess of positrons or antiprotons could indicate the presence of dark matter particles decaying or annihilating into standard particles. Experiments like the Alpha Magnetic Spectrometer (AMS) on the International Space Station and the DAMPE satellite are designed to measure cosmic ray fluxes with high precision.

Particle Colliders

Particle colliders, such as the Large Hadron Collider (LHC) at CERN, provide a controlled environment to search for dark matter candidates by recreating conditions similar to those just after the Big Bang. The idea is to produce dark matter particles through high-energy collisions of protons or heavy ions.

Missing Energy Signature

One of the main techniques used in colliders is to search for events with missing transverse energy (MET), which can indicate the presence of invisible particles like dark matter. In these events, the total momentum in the plane perpendicular to the colliding beams does not add up to zero, suggesting that some particles escaped detection.

Searches for New Particles

Colliders also allow for searches for new particles predicted by theories beyond the Standard Model, such as supersymmetry. If a new particle is discovered, it could provide a crucial link to understanding dark matter, particularly if it is stable and has the right properties to account for the observed cosmic dark matter density.

Current and Future Experiments

The hunt for dark matter involves numerous ongoing and planned experiments, employing a wide range of techniques.

Current Experiments

1. XENONnT: Located at the Gran Sasso National Laboratory in Italy, XENONnT is a direct detection experiment using liquid xenon. It aims to significantly improve sensitivity to WIMPs by detecting nuclear recoils with unprecedented precision.

2. PandaX-4T: Based in China, PandaX-4T also utilizes liquid xenon and aims to detect dark matter with high sensitivity. It focuses on reducing background noise and enhancing detection capabilities.

3. LHC Experiments: ATLAS and CMS, two major experiments at the LHC, continue to search for new particles and phenomena that could be related to dark matter. They analyze data from high-energy proton-proton collisions, looking for events with missing energy and other unusual signatures.

Future Experiments

1. LUX-ZEPLIN (LZ): Set to be one of the most sensitive direct detection experiments, LZ aims to detect WIMPs using a large liquid xenon detector. It will be located in the Sanford Underground Research Facility in the United States.

2. DARWIN: An ambitious project in Europe, DARWIN plans to build a multi-ton liquid xenon detector to search for dark matter with unprecedented sensitivity. It aims to probe WIMP-nucleon cross-sections down to the neutrino floor, where neutrino background becomes the limiting factor.

3. Cherenkov Telescope Array (CTA): As a leading gamma-ray observatory, CTA will significantly enhance the search for dark matter through indirect detection. It will provide detailed observations of potential dark matter sources, such as dwarf spheroidal galaxies and the Galactic Center.

4. Hyper-Kamiokande: A large-scale water Cherenkov detector in Japan, Hyper-Kamiokande will search for neutrinos from potential dark matter annihilation or decay events. It will complement other neutrino observatories by providing high sensitivity and resolution.

Conclusion

The search for dark matter is a multi-faceted endeavor, involving direct detection, indirect detection, and particle collider experiments. Despite significant challenges, ongoing and future experiments hold promise for detecting dark matter or at least narrowing down the possible properties of dark matter particles. The discovery of dark matter would not only solve one of the biggest mysteries in cosmology but also potentially unveil new physics beyond the Standard Model, revolutionizing our understanding of the universe.

Chapter 6: Astrophysical Evidence

Cosmic Microwave Background (CMB)

The Cosmic Microwave Background (CMB) is a faint glow of radiation that permeates the universe, originating from the hot, dense state of the early universe shortly after the Big Bang. The CMB provides a snapshot of the universe when it was just 380,000 years old, and it is a crucial source of information about the early universe's conditions and the content of the universe, including dark matter.

Anisotropies and Power Spectrum

The CMB is not perfectly uniform; it contains tiny temperature fluctuations, or anisotropies, which are imprinted with information about the early universe's density fluctuations. These anisotropies are analyzed through the power spectrum, which shows the amplitude of fluctuations at different angular scales.

1. Acoustic Peaks: The power spectrum of the CMB reveals a series of peaks and troughs, known as acoustic peaks. These peaks result from sound waves in the early universe, which were influenced by the interplay of gravity and pressure. The positions and heights of these peaks depend on the density of baryons, dark matter, and dark energy. The first peak corresponds to the largest scales of fluctuation and provides a measure of the total matter density, including dark matter. The second and third peaks provide additional information about the baryon density and the dark matter to baryon ratio.

2. Damping Tail: At smaller scales, the power spectrum shows a decline in amplitude, known as the damping tail. This is due to the diffusion of photons in the early universe, which smooths out fluctuations. The damping tail provides insights into the physics of the early universe and the properties of particles, including potential interactions with dark matter.

Polarization

In addition to temperature fluctuations, the CMB also exhibits polarization, which provides further information about the early universe. The polarization can be decomposed into E-modes and B-modes, with the E-modes being primarily generated by density fluctuations, and B-modes being associated with gravitational waves or lensing effects. The study of CMB polarization helps refine our understanding of the universe's content, including dark matter.

1. E-mode Polarization: The E-mode polarization pattern results from Thomson scattering of CMB photons by free electrons, influenced by density fluctuations. This polarization pattern provides complementary information to the temperature anisotropies, particularly regarding the ionization history of the universe and the density of different components, including dark matter.

2. B-mode Polarization: The detection of B-mode polarization, especially at large angular scales, could provide evidence for primordial gravitational waves from the inflationary period. However, at smaller scales, B-modes can arise from gravitational lensing of E-modes, offering insights into the distribution of mass in the universe, including dark matter.

Implications for Dark Matter

The CMB provides strong evidence for the existence of dark matter. The precise measurements of the CMB by experiments like the Wilkinson Microwave Anisotropy Probe (WMAP) and the Planck satellite have constrained the dark matter density in the universe. These measurements indicate that dark matter constitutes approximately 27% of the total energy density of the universe, with ordinary matter making up only about 5%.

The shape and amplitude of the acoustic peaks in the CMB power spectrum are sensitive to the amount and nature of dark matter. For instance, the presence of dark matter affects the growth of cosmic structures by providing additional gravitational pull, leading to more pronounced acoustic peaks. Moreover, the detailed analysis of the CMB anisotropies has helped rule out some alternative theories to dark matter, such as modifications to Newtonian dynamics.

Galaxy Clusters

Galaxy clusters are the largest gravitationally bound structures in the universe, containing thousands of galaxies, hot gas, and dark matter. They provide critical evidence for dark matter through several observational methods, including galaxy motions, X-ray emissions, and gravitational lensing.

Velocity Dispersion

The velocities of galaxies within clusters are too high for the clusters to remain bound solely by the gravitational pull of the visible matter. This phenomenon, known as velocity dispersion, suggests the presence of a significant amount of unseen mass, which is attributed to dark matter.

1. Virial Theorem: The virial theorem relates the kinetic energy of galaxies within a cluster to the gravitational potential energy. Observations show that the kinetic energy, inferred from galaxy velocities, is much higher than can be accounted for by the visible mass alone, implying a substantial dark matter component.

2. Mass-to-Light Ratio: The mass-to-light ratio of galaxy clusters, derived from galaxy velocities, is much higher than that of individual galaxies or the stars within them, indicating that most of the cluster's mass is dark.

X-ray Emissions and Intracluster Medium

The intracluster medium (ICM), composed of hot, X-ray-emitting gas, constitutes a significant portion of a cluster's baryonic matter. The temperature and distribution of this gas provide further evidence for dark matter.

1. Thermal Bremsstrahlung: The X-ray emissions from the ICM are primarily due to thermal bremsstrahlung, where electrons are deflected by ions, emitting radiation. The temperature of the gas, which can reach tens of millions of degrees, requires a large gravitational potential to contain it, again indicating the presence of dark matter.

2. Hydrostatic Equilibrium: By assuming that the gas is in hydrostatic equilibrium (where the pressure gradient balances the gravitational pull), astronomers can estimate the total mass of a cluster. These estimates often show that the majority of the mass is not in the form of visible matter, but rather dark matter.

Gravitational Lensing

Gravitational lensing is a powerful tool for studying dark matter in galaxy clusters. It occurs when the cluster's gravitational field bends the light from background objects, distorting and magnifying their images.

1. Strong Lensing: In cases of strong lensing, distinct arcs and multiple images of background galaxies are produced. The geometry and distortion of these images can be used to map the distribution of mass in the cluster, revealing concentrations of dark matter.

2. Weak Lensing: Weak lensing involves subtle distortions of background galaxy shapes due to the gravitational field of the cluster. By statistically analyzing these distortions, astronomers can infer the distribution and amount of dark matter. This method is especially useful for mapping the dark matter over large areas and for studying clusters that do not produce strong lensing effects.

3. Bullet Cluster and Similar Systems: The Bullet Cluster and similar systems provide compelling evidence for dark matter through lensing studies. In these systems, the gas (traced by X-rays) is separated from the galaxies and dark matter (traced by lensing) due to collisions between clusters. The dark matter and galaxies pass through each other with little interaction, while the gas is slowed down by drag forces, leading to a separation of mass and light. This separation is inconsistent with theories that modify gravity alone and strongly supports the existence of dark matter.

Large-Scale Structure

The large-scale structure of the universe, including the distribution of galaxies, clusters, and voids, provides significant evidence for dark matter. Dark matter plays a crucial role in the formation and evolution of these structures.

Cosmic Web

The universe on large scales resembles a cosmic web, with galaxies and clusters arranged along filaments, separated by vast voids. This structure is believed to have formed from the initial density fluctuations in the early universe, amplified by gravity.

1. N-body Simulations: Large-scale simulations, known as N-body simulations, are used to model the evolution of the universe under the influence of gravity. These simulations assume the presence of dark matter and show how small initial fluctuations grow over time to form the observed large-scale structure. The success of these simulations in matching observed galaxy distributions is a strong argument for the existence of dark matter.

2. Baryon Acoustic Oscillations (BAO): BAO are regular, periodic fluctuations in the density of visible matter in the universe, caused by acoustic waves in the early universe. These oscillations are imprinted in the large-scale distribution of galaxies and can be observed as a preferred separation scale. The analysis of BAO provides a "standard ruler" for cosmology, helping to measure the expansion rate of the universe and constraining the properties of dark matter.

Redshift Surveys

Redshift surveys map the three-dimensional distribution of galaxies by measuring their redshifts, which indicate their distances and velocities. These surveys provide detailed maps of the large-scale structure and have revealed patterns like the cosmic web.

1. Sloan Digital Sky Survey (SDSS): The SDSS has mapped millions of galaxies, providing a detailed picture of the large-scale structure. The data from SDSS and similar surveys help in understanding the role of dark matter in structure formation and evolution.

2. Cosmic Shear: Cosmic shear refers to the weak gravitational lensing effect caused by large-scale structures on the shapes of distant galaxies. By measuring the cosmic shear, astronomers can infer the distribution of dark matter on large scales and test models of structure formation.

Implications and Challenges

The astrophysical evidence for dark matter is compelling, yet the nature of dark matter remains one of the biggest mysteries in cosmology. The evidence from the CMB, galaxy clusters, and large-scale structure consistently points to a universe where dark matter plays a crucial role in shaping the cosmic landscape. However, despite the strong indirect evidence, direct detection of dark matter particles has not yet been achieved, leaving open questions about the specific properties and identity of dark matter.

The ongoing and future

observational and experimental efforts aim to shed more light on this elusive component of the universe. As technology and methods improve, there is hope that a more complete understanding of dark matter will emerge, potentially revealing new physics beyond the current models.

Chapter 7: Cosmological Implications

Dark matter has profound implications for our understanding of the universe. It plays a critical role in the evolution of cosmic structures, the formation of galaxies, and the dynamics of the universe on the largest scales. Moreover, the interplay between dark matter and dark energy is a central topic in modern cosmology, influencing our understanding of the universe's fate.

Evolution of the Universe

The evolution of the universe from the Big Bang to its current state is deeply influenced by dark matter. The standard model of cosmology, known as the Lambda Cold Dark Matter (ΛCDM) model, describes a universe dominated by dark matter and dark energy, with only a small fraction of ordinary matter.

Early Universe and Structure Formation

1. Initial Conditions and Inflation: The universe began with the Big Bang, followed by a period of rapid expansion known as inflation. This inflationary period smoothed out any irregularities and generated tiny density fluctuations, which served as the seeds for the formation of cosmic structures.

2. Dark Matter and Growth of Fluctuations: After inflation, the universe continued to expand and cool. Dark matter, being non-relativistic and interacting weakly with other particles, began to clump under gravity, amplifying the initial density fluctuations. This process led to the formation of dark matter halos, which later became the gravitational wells where galaxies formed.

3. Recombination and the CMB: About 380,000 years after the Big Bang, the universe cooled enough for protons and electrons to combine into neutral hydrogen atoms, a process known as recombination. This allowed photons to travel freely, creating the Cosmic Microwave Background (CMB). The CMB provides a snapshot of the universe at this early stage, with the observed anisotropies reflecting the density fluctuations that would eventually grow into galaxies and clusters.

Large-Scale Structure

1. Hierarchical Structure Formation: In the ΛCDM model, structure formation is hierarchical, meaning that small objects form first and merge to create larger structures. Dark matter halos formed early on, providing the scaffolding for galaxies and clusters. Over time, these structures merged and accreted more matter, leading to the complex cosmic web observed today.

2. Baryonic Matter and Feedback Mechanisms: While dark matter dominated the mass budget, baryonic matter (ordinary matter) played a crucial role in visible structures. Baryonic processes, such as gas cooling, star formation, and feedback from supernovae and active galactic nuclei, shaped the evolution of galaxies and their interaction with the surrounding dark matter.

Role in Galaxy Formation

Dark matter is essential for galaxy formation, providing the necessary gravitational potential for gas to cool and form stars. The distribution and properties of dark matter halos directly influence the structure and evolution of galaxies.

Dark Matter Halos

1. Halo Density Profiles: Dark matter halos are typically characterized by a density profile, which describes how the density of dark matter varies with distance from the center. The most commonly used profile is the Navarro-Frenk-White (NFW) profile, which describes a cusp-like increase in density towards the center of the halo. The properties of these halos, such as their mass and concentration, are crucial for understanding galaxy formation.

2. Subhalos and Satellite Galaxies: Within larger dark matter halos, smaller subhalos can exist, which may host satellite galaxies. These substructures are important for understanding the satellite populations around larger galaxies, such as the Milky Way, and provide insights into the processes of galaxy formation and evolution in different environments.

Baryonic Physics and Feedback

1. Gas Cooling and Star Formation: In the early universe, gas in dark matter halos could cool via radiative processes, leading to the formation of stars and galaxies. The efficiency of gas cooling and subsequent star formation is influenced by the properties of the dark matter halo, such as its mass and potential well depth.

2. Feedback Mechanisms: Feedback processes, such as supernova explosions and energy output from active galactic nuclei, can heat the surrounding gas and regulate star formation. These mechanisms are crucial for explaining the observed properties of galaxies, including their mass distribution, star formation rates, and the prevalence of dwarf galaxies.

3. Galaxy Morphology and Evolution: The interaction between baryons and dark matter influences the morphology of galaxies, including the formation of spiral and elliptical galaxies. Additionally, processes such as mergers and tidal interactions between galaxies can reshape their structure and trigger new phases of star formation or quiescence.

Dark Matter vs. Dark Energy

Dark matter and dark energy are two of the most significant components of the universe, yet they have fundamentally different roles and properties.

Nature and Characteristics

1. Dark Matter: Dark matter is a form of matter that does not emit, absorb, or reflect light, making it invisible to electromagnetic observations. It interacts primarily through gravity and is responsible for the formation and dynamics of cosmic structures. Its exact nature remains unknown, but it could consist of particles such as WIMPs (Weakly Interacting Massive Particles) or axions.

2. Dark Energy: Dark energy, on the other hand, is a mysterious form of energy that is driving the accelerated expansion of the universe. Unlike dark matter, dark energy has a repulsive effect, counteracting gravity on large scales. It is often associated with the cosmological constant (Λ) in Einstein's equations, but its true nature is still a subject of intense research.

Cosmic Expansion and Fate of the Universe

1. Cosmic Expansion: Observations of distant supernovae and the CMB suggest that the universe is expanding at an accelerating rate, driven by dark energy. This acceleration impacts the rate at which cosmic structures grow and the overall geometry of the universe.

2. Competing Forces: Dark matter and dark energy have opposing effects on the universe's evolution. While dark matter pulls structures together through gravity, dark energy pushes them apart. The balance between these forces determines the fate of the universe, whether it will continue to expand indefinitely, eventually collapse, or reach a steady state.

3. Future Observations: Upcoming astronomical surveys and observatories aim to better understand the properties of dark energy by measuring the expansion rate of the universe and the growth of cosmic structures. These studies will provide critical tests for theories of dark matter and dark energy and help refine our cosmological models.

Conclusion

The study of dark matter and its cosmological implications remains one of the most exciting and challenging areas in modern astrophysics. The evidence for dark matter from multiple independent observations is compelling, yet its exact nature eludes us. Understanding dark matter and its interplay with dark energy is essential for solving fundamental questions about the origin, structure, and ultimate fate of the universe. As research continues, the pursuit of dark matter detection and the exploration of dark energy's properties promise to unveil new insights into the cosmos, potentially revolutionizing our understanding of the universe's most enigmatic components.

Chapter 8: Challenges and Controversies

Discrepancies and Anomalies

Despite the strong case for dark matter, several discrepancies and anomalies challenge our understanding. These include issues related to the distribution of dark matter in galaxies and clusters, as well as unexpected observations that have led to questioning existing theories.

1. Missing Satellites Problem

The Missing Satellites Problem refers to the discrepancy between the number of small satellite galaxies predicted by cold dark matter (CDM) simulations and the fewer number observed around large galaxies like the Milky Way. CDM models predict hundreds of satellite galaxies, yet only a few dozen are observed. This discrepancy might suggest our models of galaxy formation are incomplete or that dark matter behaves differently on small scales.

2. Cores vs. Cusps Problem

Observations show that the density profiles of dark matter in the centers of galaxies are often flatter (core-like) than the steep (cusp-like) profiles predicted by simulations of CDM. This issue, known as the Cores vs. Cusps Problem, suggests that dark matter may have interactions not accounted for in standard models, or that baryonic processes (involving ordinary matter) significantly alter dark matter distributions.

3. Bullet Cluster and Collisional Evidence

The Bullet Cluster, an example of colliding galaxy clusters, provides evidence for dark matter through the separation of dark matter and ordinary matter during the collision. The gravitational lensing data (which maps the mass distribution) and X-ray data (which shows the gas distribution) reveal a separation between the two, supporting the existence of dark matter. However, the details of such collisions also raise questions about the properties and interactions of dark matter.

4. Dwarf Galaxy Dynamics

Dwarf galaxies often exhibit rotation curves that are difficult to reconcile with CDM predictions. These galaxies show less dark matter in their centers than expected, posing questions about the universality of dark matter's behavior or indicating potential flaws in our understanding of small-scale physics.

Competing Theories

Given these anomalies, alternative theories and modifications to existing models have been proposed. These range from adjustments to the properties of dark matter itself to more radical revisions of fundamental physics.

1. Modified Newtonian Dynamics (MOND)

MOND proposes a modification to Newton's laws of motion at low accelerations, which could explain the flat rotation curves of galaxies without invoking dark matter. While MOND can explain galaxy-scale phenomena, it struggles with larger scale structures like galaxy clusters and the cosmic microwave background (CMB) radiation.

2. Scalar-Tensor-Vector Gravity (TeVeS)

TeVeS extends MOND by incorporating additional fields, such as a scalar field and a vector field, within a relativistic framework. This theory aims to address some of MOND's limitations, particularly in explaining gravitational lensing and cosmological observations. However, TeVeS has its own challenges and has not yet matched the success of the dark matter paradigm in explaining a wide range of observations.

3. Fuzzy Dark Matter

Fuzzy Dark Matter hypothesizes that dark matter consists of ultralight particles, such as axions, which form wave-like structures on galactic scales. This model could potentially resolve the core-cusp problem by producing core-like density profiles in galaxies. However, the existence of such particles and their specific properties remain speculative and require further investigation.

4. Self-Interacting Dark Matter (SIDM)

SIDM posits that dark matter particles interact with each other through forces other than gravity. These interactions could lead to differences in dark matter distribution, potentially addressing issues like the cores vs. cusps problem and the missing satellites problem. SIDM is an area of active research, with efforts focused on understanding the nature and strength of these interactions.

Conclusion

The challenges and controversies surrounding dark matter highlight the complexities of modern cosmology. While the evidence for dark matter is robust, the exact nature of dark matter particles and their interactions remain elusive. Competing theories offer intriguing alternatives and modifications, each with its own strengths and limitations. Ongoing research aims to refine our models, test new hypotheses, and ultimately answer the fundamental questions about the universe's composition and evolution.

Chapter 9: Dark Matter in Popular Culture

Dark matter, with its mysterious and unseen nature, has fascinated both scientists and the general public. This fascination has found expression in various forms of popular culture, particularly in science fiction, and has influenced public perception and understanding of the universe.

Science Fiction

Science fiction has long been a medium for exploring scientific concepts and ideas, and dark matter is no exception. It serves as a versatile narrative element, often used to explore themes of mystery, discovery, and the unknown.

1. Dark Matter as a Plot Device

In many science fiction stories, dark matter is used as a plot device to introduce novel technologies or phenomena:

- Energy Source: Dark matter is sometimes depicted as an untapped energy source, capable of powering advanced civilizations or technologies. For instance, in some stories, dark matter is harnessed for faster-than-light travel or as a near-limitless power supply.

- Weaponization: Some narratives explore the weaponization of dark matter, using it as a tool for destruction or control. This theme often reflects fears about the misuse of advanced technology and the potential consequences of harnessing powerful but poorly understood forces.

2. Dark Matter and Extraterrestrial Life

Dark matter's elusive nature lends itself to stories involving extraterrestrial life forms or civilizations:

- Aliens and Dark Matter: In some narratives, aliens are depicted as being composed of dark matter or possessing advanced knowledge of its properties. These stories often explore themes of contact and communication, posing questions about how we would interact with life forms fundamentally different from our own.

- Intergalactic Travel: Dark matter is sometimes featured in stories about intergalactic travel, either as a medium through which travel occurs or as an obstacle that must be navigated. These tales often emphasize the vastness and mystery of the cosmos.

3. Dark Matter in Alternate Universes and Realities

Science fiction frequently uses dark matter as a gateway to explore alternate universes or realities:

- Parallel Worlds: Dark matter is often linked to the existence of parallel universes or dimensions. This concept is used to explore alternate versions of reality, where different physical laws might apply or where different histories have unfolded.

- New Physical Laws: Some stories speculate that dark matter could reveal new physical laws or forces, leading to a reevaluation of our understanding of the universe. This theme reflects the cutting-edge nature of dark matter research and the potential for revolutionary discoveries.

Public Perception

The portrayal of dark matter in popular culture has influenced public perception, shaping how people think about the universe and scientific inquiry.

1. The Mystery and Intrigue of Dark Matter

Dark matter's invisibility and unknown nature make it inherently mysterious, which is a significant part of its appeal in popular culture:

- Symbol of the Unknown: Dark matter often symbolizes the unknown or the unexplored. This symbolism resonates with audiences, tapping into a deep-seated curiosity about the cosmos and what lies beyond our current understanding.

- Narratives of Discovery: The search for dark matter is frequently portrayed as a grand quest for knowledge, akin to the exploration of uncharted territories. This narrative aligns with broader cultural themes of exploration and the human desire to uncover the secrets of the universe.

2. Educational Impact and Misconceptions

While popular culture can inspire interest in science, it can also lead to misconceptions:

- Misunderstandings of Dark Matter: Some depictions of dark matter in popular culture are scientifically inaccurate, leading to public misunderstandings. For example, dark matter is sometimes confused with antimatter or black holes, despite these being distinct concepts.

- Educational Outreach: Scientists and educators often use the public's interest in dark matter as a gateway to explain more about the universe and the scientific method. This outreach helps to correct misconceptions and provides a more accurate understanding of dark matter's role in cosmology.

3. The Role of Media and Popular Science

Media coverage and popular science publications play a crucial role in shaping public perception:

- Media Representation: News outlets and documentaries often highlight dark matter discoveries, framing them within larger narratives about the universe's mysteries. This coverage can generate public interest and support for scientific research.

- Popular Science Books and Shows: Authors and presenters in the popular science genre often use dark matter to illustrate broader scientific concepts. These works aim to make complex ideas accessible and engaging, helping to bridge the gap between scientific communities and the general public.

Conclusion

The portrayal of dark matter in popular culture reflects a broader societal fascination with the mysteries of the universe. Science fiction uses dark matter to explore themes of discovery, technology, and the unknown, while public perception is shaped by media representations and educational outreach. Together, these cultural elements contribute to a growing public interest in science and the ongoing quest to understand the cosmos. As research into dark matter continues, its role in both science and popular culture is likely to evolve, offering new opportunities for storytelling and public engagement.

Chapter 10: Conclusion and Future Directions

Summary

Dark matter remains one of the most intriguing and fundamental mysteries in modern cosmology. Its existence is supported by a broad array of observational evidence, including the rotation curves of galaxies, the dynamics of galaxy clusters, and the Cosmic Microwave Background (CMB). Despite this robust evidence, the precise nature of dark matter remains elusive, making it a central focus of current astrophysical research.

Key Findings

1. Evidence for Dark Matter: Observational data, such as the motion of galaxies and the distribution of cosmic structures, consistently support the presence of dark matter. The ΛCDM model, which incorporates dark matter, successfully explains many large-scale features of the universe.

2. Challenges and Anomalies: Discrepancies such as the missing satellites problem, the core-cusp issue, and unexpected dynamics in dwarf galaxies highlight the need for a deeper understanding of dark matter. These challenges suggest that our current models might require adjustments or extensions.

3. Theoretical Models: Various theories, including Modified Newtonian Dynamics (MOND), Scalar-Tensor-Vector Gravity (TeVeS), fuzzy dark matter, and self-interacting dark matter (SIDM), offer alternative explanations for dark matter phenomena. Each theory has its strengths and limitations, contributing to the ongoing debate in the field.

4. Technological Advancements: Improvements in observational techniques, such as more sensitive telescopes and advanced particle detectors, are providing new insights into dark matter. These advancements are crucial for testing existing theories and discovering new ones.

Unanswered Questions

Despite significant progress, several fundamental questions about dark matter remain unanswered:

1. What is the Nature of Dark Matter?

The exact nature of dark matter particles is still unknown. While WIMPs (Weakly Interacting Massive Particles) and axions are leading candidates, no definitive detection has been made. Identifying the fundamental particles that constitute dark matter is a key goal for both experimental and theoretical research.

2. How Does Dark Matter Interact?

Understanding how dark matter interacts with itself and with ordinary matter is crucial. Current models assume weak interactions, but alternative theories propose different interaction mechanisms. Investigating these interactions could reveal new physics or necessitate modifications to existing models.

3. What Role Does Dark Matter Play in Structure Formation?

The role of dark matter in the formation and evolution of cosmic structures is well-established, but the details remain complex. Questions about the effects of dark matter on galaxy formation, halo structure, and baryonic processes are still actively researched.

4. How Does Dark Matter Relate to Dark Energy?

The relationship between dark matter and dark energy is a significant area of inquiry. While dark matter contributes to the gravitational pull that forms structures, dark energy is responsible for the accelerated expansion of the universe. Understanding how these components interact and influence each other is crucial for a comprehensive cosmological model.

Future Research

Future research into dark matter will involve a combination of observational, experimental, and theoretical efforts:

1. Direct Detection Experiments

Direct detection experiments aim to observe dark matter particles interacting with ordinary matter. New-generation experiments, such as those using ultra-sensitive detectors and underground laboratories, are designed to detect potential dark matter interactions. Continued refinement of detection methods and the exploration of different types of detectors will be essential for making progress in this area.

2. Particle Physics Experiments

Particle physics experiments, particularly those conducted at high-energy colliders like the Large Hadron Collider (LHC), may produce dark matter particles or provide evidence for new physics beyond the Standard Model. These experiments can test theoretical predictions and search for signatures of dark matter interactions.

3. Astrophysical Observations

Astrophysical surveys and observations are crucial for understanding dark matter's role in the cosmos. Upcoming missions, such as the James Webb Space Telescope (JWST) and the Euclid satellite, are expected to provide new insights into dark matter's distribution, structure, and influence on cosmic evolution. Detailed observations of galaxy clusters, cosmic microwave background fluctuations, and large-scale structure will continue to refine our understanding.

4. Theoretical Developments

Theoretical physicists are exploring new models and extensions to existing theories to better explain dark matter phenomena. Research into alternative theories, such as modified gravity or quantum gravity models, may provide new insights into dark matter's nature and interactions. Cross-disciplinary approaches combining theoretical physics with astrophysical and experimental data will be crucial for advancing our knowledge.

5. Interdisciplinary Collaboration

Collaboration between astrophysicists, particle physicists, and cosmologists will be vital for addressing the complex questions surrounding dark matter. Interdisciplinary research can lead to innovative approaches and comprehensive models that integrate observations, experiments, and theoretical insights.

Conclusion

The quest to understand dark matter is a major scientific endeavor with profound implications for our knowledge of the universe. While significant progress has been made, many questions remain. The future of dark matter research promises to be dynamic and transformative, with new technologies, theories, and collaborative efforts poised to make groundbreaking discoveries. As we advance in our quest to unveil the secrets of dark matter, we will not only deepen our understanding of the cosmos but also explore the fundamental nature of reality itself.

References

Books

1. **"Dark Matter and the Dinosaurs: The Astounding Interconnectedness of the Universe"** By Lisa Randall HarperCollins, 2015.

2. **"The Cosmic Cocktail: Three Parts Dark Matter"** By Jennifer A. Johnson Princeton University Press, 2020.

3. **"Introduction to Modern Cosmology"** By Andrew Liddle Wiley, 2015.

Online Resources

1. **NASA's Dark Matter Page**
   https://www.nasa.gov/mission_pages/dark_matter/index.html

2. **European Space Agency's Dark Matter Research**
   https://www.esa.int/Science_Exploration/Space_Science/Dark_Matter_Research_ESA_Overview

   CERN's Dark Matter Research Overview
   https://home.cern/science/physics/dark-matter

Glossary

Key Terms

- Dark Matter: A form of matter that does not emit, absorb, or reflect light, making it invisible. It is thought to constitute about 27% of the universe's mass and energy.

- WIMP (Weakly Interacting Massive Particle): A hypothetical dark matter particle that interacts via the weak nuclear force and gravity. WIMPs are among the leading candidates for dark matter.

- Axion: A theoretical elementary particle proposed as a candidate for dark matter. Axions are very light and interact very weakly with other matter.

- Cosmic Microwave Background (CMB): Radiation left over from the early hot phases of the universe, providing critical evidence for the Big Bang theory and insights into dark matter.

- Modified Newtonian Dynamics (MOND): A theory proposing modifications to Newton's laws to account for the observed rotation curves of galaxies without invoking dark matter.

- Scalar-Tensor-Vector Gravity (TeVeS): A relativistic theory extending MOND By incorporating additional fields to explain dark matter phenomena.

- Bullet Cluster: A pair of colliding galaxy clusters whose observation provides strong evidence for the existence of dark matter.

Concepts

- Dark Energy: A mysterious form of energy that is causing the accelerated expansion of the universe. It is distinct from dark matter but plays a significant role in cosmological models.

- Rotation Curves: Graphs showing how the rotational speed of stars or gas in a galaxy changes with distance from the galaxy's center. These curves provide evidence for dark matter in galaxies.

- Galaxy Cluster: A large structure consisting of hundreds to thousands of galaxies bound together By gravity. Observations of galaxy clusters help in studying dark matter and its distribution.

www.ingramcontent.com/pod-product-compliance
Lightning Source LLC
Chambersburg PA
CBHW072002210526
45479CB00003B/1035